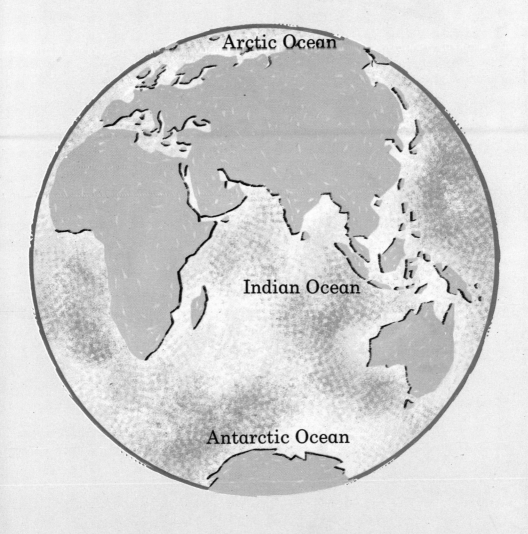

First Published in 1970 by
Macdonald and Company
(Publishers) Limited
St. Giles House
49-50 Poland Street
London W1

Managing Editor
Michael W. Dempsey B.A.

Chief Editor
Angela Sheehan B.A.

Made and printed in Great Britain
by A. Wheaton & Company
Exeter Devon

MACDONALD FIRST LIBRARY

Water

Macdonald Educational
49-50 Poland Street
London W1

Most of the world is covered by sea.
The water in the sea is salty.

There is also a lot of water on the land.
There is water in the soil and in streams.
This water comes from rain.
It is not salty.

All plants and animals need water.
The plants and animals which live in the sea
have water all around them.
But the animals and plants which live on the
land cannot drink salty sea-water.

The animals drink water from ponds and
streams.
The plants take in water from the ground
through their roots.

3

well

When it rains, some of the water soaks into the ground.
If you dig deep enough you will find water.
Some people get water from deep holes in the ground.
These holes are called wells.

spring

In some places the underground water rises to the surface, as a spring.

Rivers start from springs.

Rivers start as little
trickles of water.
They get bigger and
bigger as more and more
water flows into them.

The water in a river is
moving all the time.
It always flows downhill
towards the sea.

river

Lakes are made when
water fills hollows in
the ground.
Some lakes are so big
that you cannot see the
other side.
Very small lakes are
called ponds.

lake

5

Most people get their water from taps.
The water has to travel a long way to reach
the taps.
The water we use is pumped from a river and
kept in a lake, called a reservoir.
Water is stored in the reservoir until it is
needed.
Then the water goes down a pipe to the
waterworks where it is made clean.

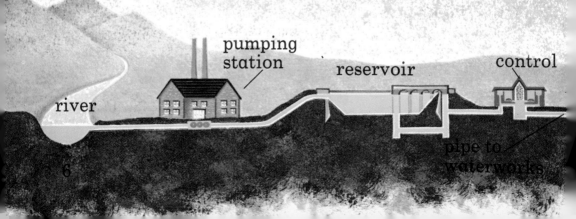

pumping
station

reservoir

control

river

pipe to
waterworks

The water soaks through layers of gravel and
sand.
The dirt stays behind in the sand.
Then a chemical, called chlorine, is
put into the water to kill any germs.

The clean water is stored in a big tank.
Then it goes along a big pipe under the road.
This pipe is called a water-main.

Smaller pipes take the water to each house.

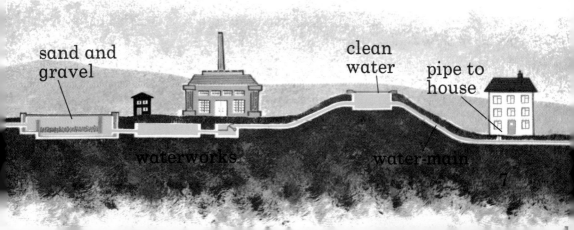

sand and
gravel

clean
water

pipe to
house

waterworks

water-main

7

rain-water
drain

drain

sewer

The dirty water from a house cannot go
straight back into the rivers.
The soap in the water would kill the fishes.
The dirt would spread diseases.
The dirty water goes down a pipe to the drain.
Then the water flows through a big pipe,
called a sewer, to the sewage-works.

8

At the sewage-works the dirty water is put
into big tanks.
Most of the dirt falls to the bottom.
Then the water soaks through cinders.
The rest of the dirt is left in the cinders.
Then the water is clean enough to go into a
river, but it is not clean enough to drink.

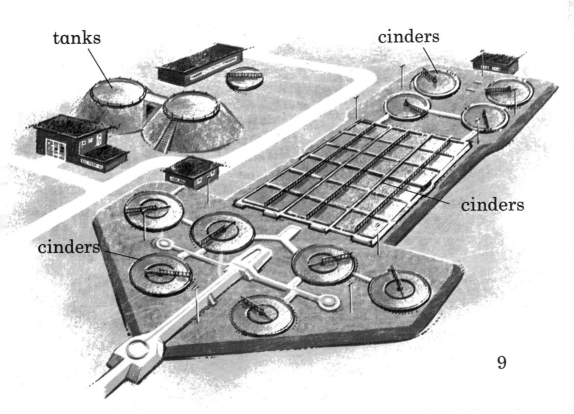

tanks

cinders

cinders

cinders

Without water the world would be a very dirty place.

We use water all the time to make things clean.

We use lots of hot water and soap when we have a bath.

The soap helps to loosen the dirt.

The dirt floats away in the water.

Many things must be washed in a home.

Clothes and dishes get dirty.

Walls and floors must be cleaned.

All the dirt must be washed away.

There are germs in dirt which can make us ill.

Things look better when they are clean.

Cars are washed to make the paint shine.

Some people use water to warm their houses.
The water is heated in a boiler.
The hot water from the boiler goes along a
pipe to radiators.
A radiator is like a metal hot-water bottle.
The heat of the radiator warms the air.
When the water has been through the
radiators it goes back to the boiler to be
heated again.

Cold water is used to keep things cool.
When a car engine is working it gets very hot.

A car radiator is filled with cold water,
to keep the engine cool.
Pipes from the radiator carry cold water
round the engine.
When the water has been round the engine
it is quite hot.
It goes back to the radiator to cool.

Water is used to work machines which make electricity.
The water is stored in a lake behind a big wall.
The wall is called a dam.

The water rushes through a pipe at the bottom of the dam.
The water goes so fast that it makes a big wheel spin round and round.
The wheel is called a turbine.

The turbine drives a machine called a generator.
The generator makes electricity.

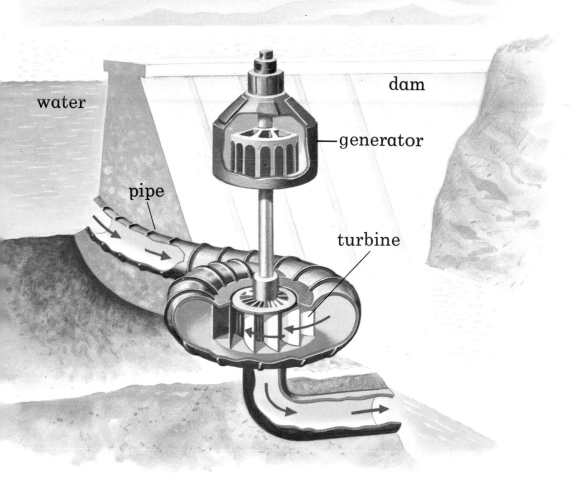

water

dam

generator

pipe

turbine

cargo ship

passenger ship

Ships float in water.
They can cross the big oceans.
Ships carry people and cargoes from one
country to another.
A lot of our food is brought in ships from
faraway lands.

Ships also travel on rivers.
Sometimes 'rivers' are built between big towns.
They are called canals.
Special boats, called barges, travel on canals.
They are long and narrow.
They carry heavy things, such as coal.

If you look into a pond you will see a
picture of yourself in the water.
The picture is called a reflection.
It is like the reflection you see in a mirror.

Water reflects the colour of the sky.
When the sky is blue the water is blue.
When the sky is cloudy the water is grey.

You can see through a glass of water.
Glass and water both let through light.
They are transparent.
But things look different in water.
If you put a paint brush in water it looks
bent.
Everything seems to bend as it goes into
water.

When it is very cold water freezes.
It turns to ice.

In winter, water in a pond may freeze.
When the ice is thick enough, people can
walk and skate on the pond.
When the weather gets warmer the ice melts.
It turns back to water.

Ice is lighter than water.
If you put a piece of ice into a glass of
water the ice will float.

In cold parts of the world big lumps of ice
float in the sea.
They are called icebergs.
Some icebergs are bigger than houses.

When wet clothes are hung on a line, the sun
soon dries them.
The water turns to vapour and disappears
into the air.
The clothes will dry on a cloudy day.
But they dry much faster when the sun is
shining.
Heat makes water turn to vapour much faster.
The sun shining on the sea turns a lot of
water to vapour all the time.

When the vapour cools it turns back to water.
Vapour from a boiling kettle turns back to
little drops of water which float in the
air.
The drops of water are called steam.
When vapour cools high in the sky it makes
clouds.
Clouds are just like steam.

Clouds are made of millions of little drops
of water.
The drops are so small and so light that
they stay up in the air.
Sometimes the tiny drops join together and
make big drops.
These drops are too heavy to float in the air.
They fall to the ground as rain.

When the weather is very cold the
tiny drops of water in the clouds
turn to ice.

Each little piece of ice is a crystal.
The ice crystals grow bigger and fall
as snowflakes.

Snowflakes are very beautiful.
Each snowflake has six points
and each one has a different pattern.

As the sun shines on the sea, lots of water turns to vapour and goes into the air.
As the air moves over the land it may have to rise over hills and mountains.
The higher the air goes, the cooler it gets.
Some of the vapour turns to water.

The little water drops form clouds in the sky.
If the water drops grow bigger they fall to
the ground as rain.
Most of the rain that falls runs into rivers
which flow to the sea.

In some hot lands it rains nearly every day.
Big forests, called jungles, grow in these
places.

The rain makes the trees grow very fast and
very tall.

Climbing plants grow up the tree trunks to reach the sunlight.
Many animals live in jungles.
There are snakes on the ground.
There are birds and monkeys in the trees, and insects buzzing everywhere.

Deserts are places where
it hardly ever rains.
Only special plants can
live in deserts.
Cactus plants store
water in their fat stems.
They have sharp prickles
to stop animals eating
them.

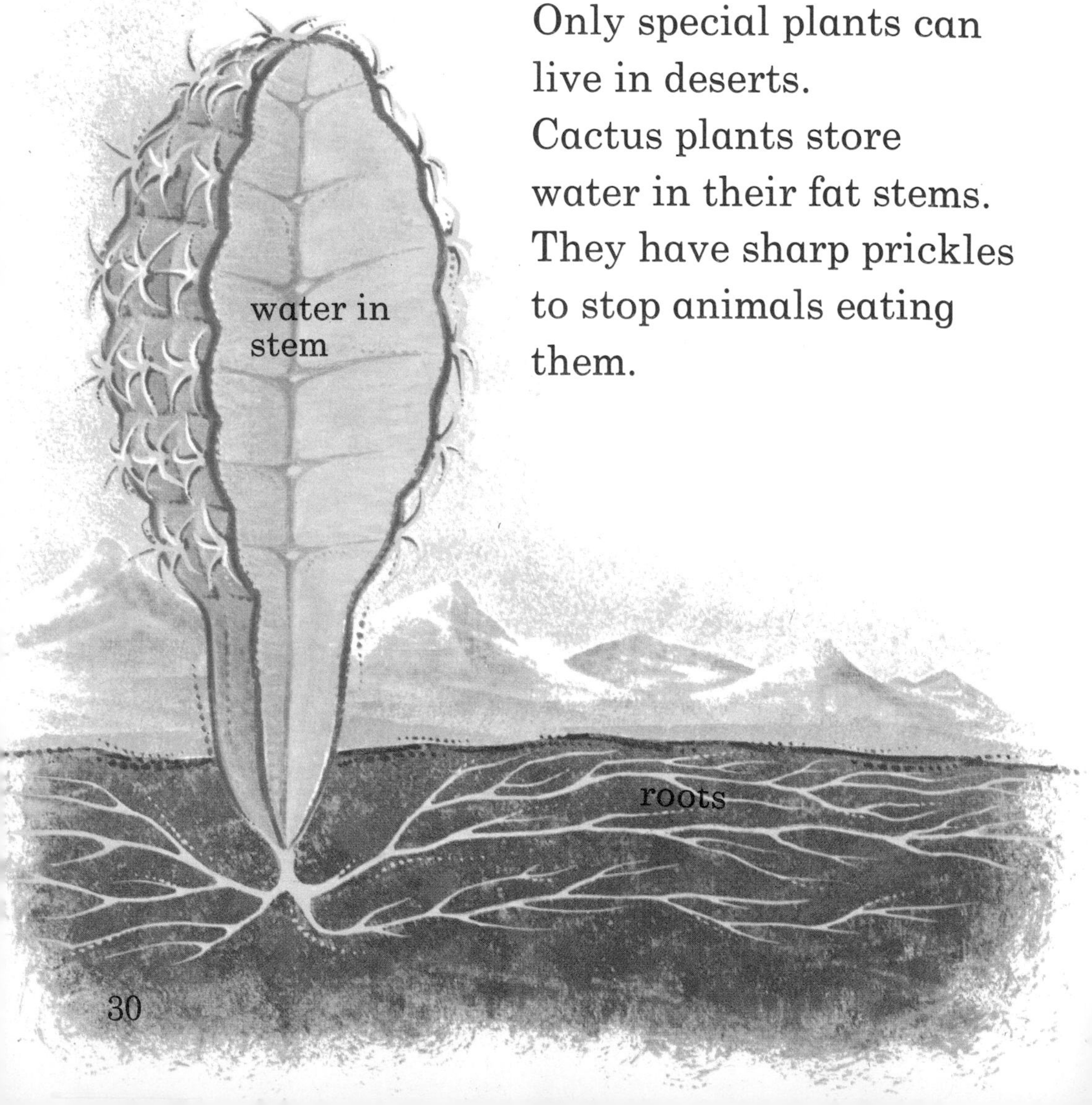

water in
stem

roots

In dry lands farmers use water from rivers
to make their crops grow.
The water flows in ditches through the
fields and makes the ground wet.
Long ago some farmers used a bucket on the
end of a pole to get the water from rivers.
A stone on the other end of the pole made it
easy to lift the bucket.
Now water is pumped from rivers to the fields.

Index

MACDONALD FIRST LIBRARY

Oceans of the world

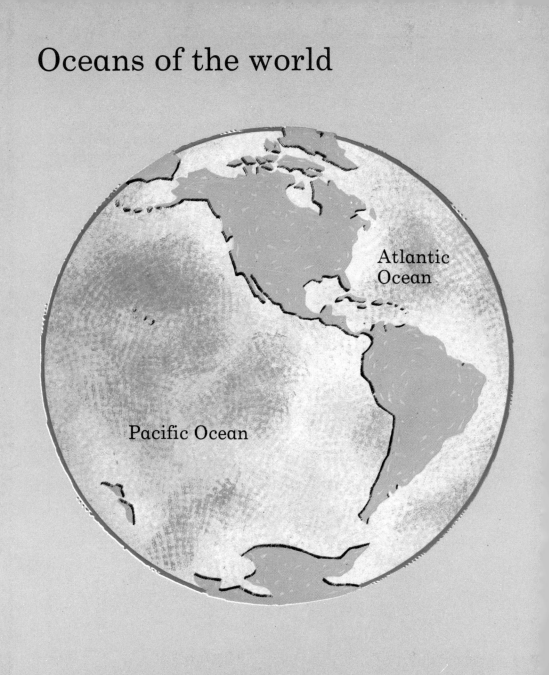

Atlantic
Ocean

Pacific Ocean